W9-AQW-191

The Wild Blueberry BOOK

The Wild Blueberry BOOK

VIRGINIA M. WRIGHT

Down East

ISBN: 978-0-89272-939-5

Design by Rich Eastman
Cover Design by Miroslaw Jurek
Front cover photograph by Stacey Cramp

Library of Congress Cataloging-in-Publication Data

Wright, Virginia (Virginia M.)
The wild blueberry book / by Virginia Wright.
p. cm.
ISBN 978-0-89272-939-5
1. Blueberries--Popular works. 2. Blueberry industry--Maine--Popular works. 3. Cooking (Blueberries)--Popular works. I. Title.
SB386.B7W75 2011
634'.737--dc22

2011007492

Printed in China
5 4 3 2 1

Distributed to the trade
by National Book Network

Contents

The Wild Bunch

She was a New Yorker on her first trip to Maine, and I couldn't help but eavesdrop as she perused the produce at a midcoast farm stand. In her hands she cradled a green quart box brimming with dusky blueberries. "Ooooo, look at these baby berries!" she cooed, gently nudging the fruit. "Look how tiny they are! Aren't they sweet?"

Sweet they are, I wanted to tell her, and tart, too, but they are not, as she presumed, "baby," or immature, versions of the familiar marble-sized blueberries grown in New Jersey, Michigan, and other places around the world. They are *wild* blueberries, an ancient species (*Vaccinium angustifolium*) that took root in North America as the glaciers receded ten thousand years ago and thrived in the sandy, acidic soils that other plants found inhospitable. Naturally growing stands of wild, or lowbush, blueberries are commercially harvested only in Maine and eastern Canada, and fresh ones are almost impossible to find in markets south of New England.

Grazing on them in roadside patches and on granite hilltops is a mid-summer rite in these parts. Each berry bursts with its own flavor. The first may be sweet, the second tart. One may be grapey, another citrusy. Some are intense. Others are subtle. As you pluck your way through a mass of the low-growing shrubs, you begin to associate

Jennifer Smith-Mayo

tastes with colors: maybe the indigo ones are sugary, the black ones piquant, the turquoise ones tangy. Pop a handful into your mouth and you get a jammy fusion of flavors (purple lips and purple teeth, too). Pour a couple of cups of them into a piecrust, and the filling bakes and oozes into something deep, intricate, and just plain wonderful.

There's a reason wild blueberries from one bush taste different from those of its neighbor: they *are* different. One acre of wild blueberries typically contains well over one hundred varieties of the berry, each one as genetically distinct from the other as a McIntosh apple is from a Delicious. The shin-high plants mingle as they grow, spreading underground by way of a shallow rhizome root system that sprouts new shrubs, which are clones of the original bush (some very old clones sprawl over areas as large as football fields). "That rich genetic diversity is what gives wild blueberries their unique flavor," says David Yarborough, a wild blueberry specialist with the University of Maine Cooperative Extension. "You also get a mix of overripe and underripe fruit along with fruit that's perfectly mature, which adds to the complexity of their taste." He estimates that there are 6.5 million distinct wild blueberry clones in Maine alone. And since the bushes cross-pollinate, there always is potential for new clones to be created.

By contrast, there are just slightly more than one hundred varieties of the cultivated, or highbush, blueberry species, and since a grower is likely to plant only one or two of them, his field yields berries that are uniform in size, color, and flavor. Larger, sturdier, and easier to pick than wild blueberries, highbush berries are the variety you're most likely to find in the supermarket produce section. But if you relish fruit for its flavor, not ease of harvest, you'll find the cultivated globes, while tasty, can't compete with the tart, distinctive flavor of wild blueberries.

Why Do They Call Them "Wild"?

Lowbush blueberries are called "wild" because, unlike domesticated highbush blueberry bushes, they are not planted from seeds or cuttings and their millions of varieties were created by bees transferring pollen, not by researchers wielding tweezers. Commercially-owned naturally-growing stands of wild blueberries get plenty of cultivation though, coaxed as they are to maximum production from the releasing of those bees in spring to the mowing or burning of fields to prune shrubs in fall.

Cultivated Blueberries

Cultivated, or highbush, blueberries are in the same genus—*Vaccinium*—as wild, or lowbush, blueberries, but they are different species. The three most common types of highbush blueberries are:

Northern Highbush Blueberry (***Vaccinium corymbosum***) Domesticated from native North American plants in the 1920s, the northern highbush blueberry is a youngster compared to wild blueberries, which are early colonizers dating to the end of the last ice age. Plants typically grow four to seven feet tall depending on the cultivar.

Rabbiteye Blueberry (*Vaccinium ashei* or *Vaccinium virgatum*). Native to the southeastern United States, these berries are pink (like a rabbit's eyes) before they turn blue. They are sweeter than northern highbush blueberries, but their skin is a little tougher after freezing. They grow taller, too – as much as twenty feet if not pruned.

Southern Highbush Blueberry (*Vaccinium corymbosum hybrid*). This blueberry is a relatively new heat-tolerant cross between the northern highbush blueberry and native southern species.

Savory Salad with Goat Cheese and Wild Blueberry Sauce

From *Wild Blueberry Association of North America.*

- 2 large or 3 medium shallots
- 2 tablespoons olive oil
- 1½ cups frozen wild blueberries
- 3 ounces water
- 3 tablespoons Grey Poupon mustard
- 1 tablespoon peach or apricot preserves
- 1 tablespoon cornstarch
- ½ teaspoon salt
- 2 endives
- 1 radicchio
- 2 yellow peppers
- 2 tablespoons white wine vinegar
- ¼ teaspoon salt
- ⅛ teaspoon pepper or more to taste
- 1 pinch sugar
- 3 tablespoons olive oil
- ½ teaspoon ground coriander
- Six 2-inch rounds of goat cheese ½-inch thick (about 6 ounces)
- 2 tablespoons powdered sugar

Sauce: Peel and dice shallots. Sauté in 2 tablespoons of olive oil until softened about 3 to 5 minutes. Stir in berries, 3 ounces of water, mustard, and preserves. Simmer 3 to 5 minutes. Mix cornstarch with a little cold water until the mixture is smooth. Add to the blueberry mixture, stirring well. Bring to a boil, cook for 2 to 3 minutes. Add salt. Let cool slightly.

Salad: Clean and wash endive, radicchio, and peppers. Cut endive into bite-size cubes. Slice yellow peppers into thin strips. Cut radicchio into bite-sized pieces. Mix vinegar, salt, pepper, sugar, and olive oil in big bowl. Add salad ingredients and toss well.

Cheese: Dredge top of goat cheese rounds in powdered sugar. Place goat cheese on baking tray covered with foil. Brown slightly under preheated grill, 1 to 2 minutes. Remove and sprinkle with coriander. Preparation time: approximately 35 minutes.

Did You Know?

- The blueberry is part of the *Ericaceae* family of plants, which includes heaths, heathers, rhododendrons, azaleas, and Indian pipes.

- Each wild blueberry contains about 50 seeds.

- The blueberry rake was invented by a Down East man, Abijah Tabbut, in 1822.

- Maine produces about 38 percent of the world's wild blueberries and 15 percent of all blueberries, wild and domesticated, in North America.

Blueberries by the numbers...

90: number of domesticated blueberries in 1 cup.

150: number of wild blueberries in 1 cup.

15: percentage of North American blueberries, both domesticated and wild, grown in Maine.

1: percentage of the wild blueberry crop that is sold fresh.

Other BLUE Berries

Bilberry. The bilberry is closely related to the blueberry (it is sometimes called the European blueberry), but it is smaller and blue throughout the pulp, a visible sign that it is even richer in anthocyanins, pigments that are high in health-promoting antioxidants. Bilberries, which are quite sour, are commonly consumed in Europe as a "nutraceutical," as foods that provide health benefits are called. In fact, World War II Royal Air Force pilots ate bilberry jam to improve their night vision. Whortleberry, blaeberry, whinberry are some of the many colloquial names for the bilberry.

Huckleberry. What folks in the western United States call a "huckleberry" is actually a bilberry and part of the *Vaccinium* genus, which also includes the blueberry. Depending on the species, the berries may be black, purple, red, or bright blue. The eastern huckleberry is an entirely different shrub belonging to the genus *Gaylussacia*. It produces a glossy black fruit. Both types of huckleberries are edible.

Jennifer Smith-Mayo

Blue Acres

The annual ritual begins when soft blue shadows spread over the vast barrens of Down East Maine in early August. Hundreds of men and women — local folks, eastern Canadian Mi'kmaqs, and migrants from Mexico, Honduras, and Haiti — fan out over the fields. They stoop and swing short-handled rakes through the blueberry shrubs, breaking their cadence now and then to spill their harvest into brightly colored plastic crates.

As these men and women labor under the midsummer sun, pickup trucks poke and idle on the rough dirt tracks that lace the barrens. Men trail behind them, collecting filled crates and stacking them high on the truck beds for their journey to the factory, where most of the berries will be frozen.

"You won't find this sight anywhere else in the country," boasts Del Emerson, a blueberry farmer in the town of Addison and the retired manager of Blueberry Hill Farm in Jonesboro, the University of Maine's wild blueberry research facility. "This is the only place in the U.S. where wild blueberries grow on this scale."

That scale is a relatively new phenomenon in an agrarian tradition that stretches to pre-colonial America. The Wabanaki Indians harvested "star berries" — so named for the star-shaped remnant of a flower calyx on each blueberry's crown — to eat fresh, to add to stews, and to dry for an ingredient of pemmican, a paste of dried meat and fat that was a staple of their diet. They recognized the blueberries' health benefits, brewing wild blueberry tea to relieve stomach aches and boiling the juice into thick syrup to treat sore throats and coughs.

The Indians innovated one of the idiosyncrasies of wild blueberry farming: burning the fields to promote growth. "Scientists have done core samples in bog areas where they've found fire layers that are thousands of years old, evidence that the Indians were managing blueberries very early on," says University of Maine wild blueberry researcher David Yarborough. "They knew that as the forest grew up, they got less production from the blueberries, and that lightening strikes and other burned areas opened up the barrens." Rather than wait for a lightning strike, the Indians deliberately set the fields afire every few years. They taught that management technique to European settlers, and it is still widely used.

The rolling, rugged barrens of Washington County, where the sun first shines on the United States each morning, comprise about half of Maine's commercially harvested wild blueberry acreage. Another 11,000 acres of blueberries are spread among the slopes of Hancock County on Maine's midcoast, and the rest are scattered through the state.

Jennifer Smith-Mayo

One cup of blueberries contains 83 calories, 0.5 grams of fat, 0 mg cholesterol, 21 grams of carbohydrate, 3.5 grams of fiber, and 1.1 grams of protein.

Big Berry!

The largest edible blueberry ever recorded weighed eight grams (nearly one-quarter of an ounce) and was nearly an inch in diameter. Sixteen-year-old Zachary Wightman, whose family has a blueberry farm in Kerhonkson, New York, presented the highbush blueberry for weighing and exhibition at the Ulster County Fair in 2009.

Together these lands yield about 76 million pounds of wild blueberries a year, more than any other place in the world. For that, Blueberry Hill Farm gets much of the credit.

The research farm was founded by an act of the Maine State Legislature in 1945, when growers were harvesting about thirteen million pounds of blueberries from 30,000 acres annually, and its work has been decidedly, well, fruitful. "We developed weed killers to control the grasses and woody plants," says Emerson, who managed the farm for thirty-five years. "With the weeds controlled, we were able to use fertilizers — before that, fertilizers just fed the weeds. The blueberries flourished. We worked on increasing the plant stand — that is, we'd add mulch to bare spots and the blueberries would grow into that area. And we started using honeybees to pollinate the plants. And not just native bees. We bring in millions and millions of bees each spring."

Now totaling sixty acres, Blueberry Hill Farm continues its blueberry research and experiments in Jonesboro and at private farms around the state.

Ike Hubbard and his two-handled rake.

Raking It In

As new farming techniques improved blueberry fields and barrens in the late 1980s, they presented an interesting problem: the soft tin rakes that workers had used for decades couldn't stand up to the vigorous shrubs. Handles broke, leaving rakers berryless and frustrated. For help, they turned to the neighbor with the mechanical engineering degree, Ike Hubbard.

It was serendipitous for Hubbard, who had recently given up a decades-long career in engineering and sales in Portland in order to return to his native Jonesport (he couldn't bear to see his late uncle's house sold to someone "from away," he explains, so he bought it). "The workers said, 'Ike, we need a rake that won't break,'" he says, so he fashioned a few of the tools with aluminum baskets and spring steel teeth, and the Hubbard Rake Company was born.

Working in his uncle's machine and metal shop, Hubbard makes several styles and sizes of rakes, including one with an extended handle that he originally designed for a friend who had "acquired a bigger gut" over the winter. He's best known for the two-handled rakes that he developed with the help of Mexican migrant workers as part of a study into carpal tunnel syndrome. "The two handles allow you to balance the weight with both hands when you're working in extra thick bushes," Hubbard says.

Blueberry Raker Brian Francis

Each summer, hundreds of Canadian Mi'kmaqs travel south to Columbia Falls, Maine, to rake for Notheastern Blueberry Company, which is owned by Maine's Passamaquoddy Tribe. Among the rakers is Brian Francis, a resident of Big Cove First Nation, a Mi'kmaq community of 2,484 people in New Brunswick.

How long have you been raking blueberries?
I've been going to Maine at the end of July or early August as long as I can remember — ever since I was a child. I'm fifty now. My father used to have a crew in Sedgwick, near Blue Hill. We worked for a number of leaseholders there. In 1978 we migrated up to Columbia Falls.

Do the people from Big Cove travel together as crews?
People come as families. They load up their vans and off they go. In August, our whole town practically shuts down because everyone is down to Maine. One of the parish priests even went to do mass on the blueberry fields because that's where his congregation was. Recently some of our people have gotten into fishing, so a lot of people who would have gone to Maine in the past now stay home to fish.

Is raking on the barrens in Columbia Falls different than raking the fields on the midcoast?
In Sedgwick, there were all these boulders in our way. It wasn't an easy place to rake. We had always heard about the flat fields and bumper crops around Cherryfield and Columbia Falls, and when we got to there it was like heaven. In Sedgwick, a good raker would average maybe thirty or forty boxes a day, whereas in Columbia Falls, it's easy to average one hundred boxes. [Rakers are paid $2.25-$2.50 per box. A box holds twenty-five pounds of berries.]

What's your average?
I average seventy-five boxes. My best day was one hundred-and-four boxes.

Do you have a special technique?
Just steady. A lot of times you see people taking breaks for an hour or so. Those are the ones who aren't that productive. As long as you're steady, you do well, and if you start early enough, you might reach your goal by 2 o'clock, and you can call it a day.

Does your back hurt after a day of raking?
Every muscle in your body hurts! On the first day, you tell yourself you're not going to work too hard so you won't be sore, but you always overdo it and you always get up the next day hurting all over.

Do you like it?
I love it. My love for it has a lot to do with nostalgia — I've been doing this since I was a kid. I love the social aspect of going to Maine and being in the camps. After hours, people gather around the campfire and have coffee and talk. I think it's part of the inherent nomadic instincts of a native person.

What are the camps like?
The camps are in the middle of the barrens. They recently had hot water and showers installed. Back in the 1970s, we didn't have those conveniences. They were just shacks. But we liked it. It was a real adventure. We brought a Coleman stove and bathed in a brook. We still stay in shacks, but they're fancier shacks.

Is it the same crowd at the camp year after year?
Pretty much. I bring my family, and the people my wife works with bring their families. That's one of the unique aspects: everybody takes their holidays to go work in Maine.

You mean they have full-time jobs and this is their vacation? Is that true for you, too?
Yes. I'm a television film producer. My wife is a social worker.

Really? Raking is what you do for fun?
Yeah, it is fun. And it kind of pays for the vacation.

Do your kids rake?
They just turned old enough to rake — our daughter is fifteen, and our son is twelve. They didn't like it very much.

Did you like it when you were a kid?
Not at first. I remember the real hot days. I would follow my father around and sit down in his shadow for the shade.

Is mechanization threatening this way of life?
It is. I won't say it's imminent, but it's coming. In 2010 the company brought in two mechanical harvesters for the first time, which left us with two fewer fields to do. So we do worry about it. Raking has become a tradition for our people. We all feel we belong in Maine, that we're Mainers also. We spend so much time there. The fondest memories we have are there.

The Burning Bush

Out West, wheat fields are burned after the harvest to dispose of the chaff and destroy unwanted insects, weeds, and seeds, but you'd be hard pressed to name a food crop other than wild blueberries that is burned to promote its growth. It seems counterintuitive — how can reducing a plant to charred stubble be good for it?

"Most plants get less productive as they grow old," wild blueberry researcher David Yarborough explains. "You get branching and self-shadowing and they get less productive over time. Even highbush blueberries get less productive, so growers go in and take out the big stems and open the canopy to light. But you can't do that with lowbush blueberries because it's a rhizome system that grows out, rather than up. You want to create a situation where they have maximum light because two-thirds of the plant is underground. So when you burn the plant back to the ground, you're only cutting one-third of the plant."

Burning is, in other words, a pruning technique, and it is generally performed every two years. The season immediately following a burn is a growth year in which no fruit is produced, so growers rotate fallow and fruit-bearing years among their fields. "You may think that's a bad thing — you need twice as much land to have a

David Yarborough

A crew uses a burner to burn a blueberry field.

crop every year — but in fact it's a very good thing," Yarborough says. "The insect cycle is disrupted, the diseases that affect the flower have no flower to infect, and the maggot flies have no fruit to lay their eggs in."

Even the many growers who opt for a biennial mowing of improved fields (land that has been leveled and cleared of rocks) swear by a good scorching every few years. "It's built-in pest management that wild blueberries have and highbush blueberries don't," Yarborough says.

Burned fields at Intervale Farm in Cherryfield.

Wild Blueberry Harvest

The annual wild blueberry harvest varies, sometimes dramatically, depending on temperatures, rainfall, pests, and other factors. Following are the five-year (2005-2009) averages in millions of pounds for wild blueberry production in North America.

Maine 76.99
Quebec 59.92
Nova Scotia 30.8
New Brunswick 26.56
Newfoundland 0.82
Prince Edward Island 8.96
Total: 204.05 million pounds

Source: University of Maine Cooperative Extension

Who's Who in Maine Blueberries

Wild Blueberry Commission of Maine

Members are blueberry growers and processors appointed by the governor. The commission administers Maine's wild blueberry tax, which was established in the 1940s at the request of growers and processors to raise money for research and promotion. Each pound of blueberries is taxed 1½ cents. Half is paid by growers, and half is paid by processors and shippers.

Wild Blueberry Advisory Committee

Members are blueberry growers and processors who advise the Wild Blueberry Commission on funding priorities.

Wild Blueberry Association of North America

This organization, formed in 1981, promotes wild blueberry products in the United States and Canada.

Amy Wilton

Brain Food —and More

Breakfast to Mary Ellen Camire often means a blueberry smoothie — one cup of frozen fruit whirred in a blender with skim milk, whey protein powder, and plain yogurt. She knows better than most how good it is for her: she is among the researchers whose work has established blueberries as one of the "superfoods" shown to improve health by lowering one's cholesterol, for example, or by protecting against certain cancers.

A professor of food science and human nutrition and the director of the consumer testing center at the University of Maine in Orono, Camire studies everything from how nutrition labels influence what people eat to consumer attitudes toward foods like fresh potatoes in light of the increasing popularity of frozen and processed products. When it comes to blueberries, she is specifically interested in the fruit's effect on human health. She and her UMO colleague, Dorothy Klimis-Zacas, a professor of clinical nutrition, are among roughly fifteen researchers from across the United States and Canada who convene every August in Bar Harbor for the Wild Blueberry Research Summit, where they share their recent findings in the fields of neuroscience, aging, cardiovascular disease, cancer, ophthalmology, and other areas of health.

The Lassen children graze on organic blueberries at Intervale Farm.

So what's so super about blueberries? For starters, studies suggest they can reduce the risk of heart disease by lowering cholesterol levels. They protect against urinary tract infections, thanks to a compound that inhibits growth of certain bacteria. They protect your eyes against diseases like macular degeneration because they contain antioxidants and vitamins that are essential to ocular health. And if that doesn't impress you, how's this: wild blueberries have even been shown to reverse some of the effects of aging.

The secret, according to Camire, is in that deep blue skin. "Blueberries have a lot of different kinds of health-promoting compounds," she explains. "It's those that give the berries their color — the anthocyanins — that we think are the biggest contributors. There's a whole family of these compounds that vary slightly in their chemical structure. Some are better for vascular health; others are better antioxidants. Blueberries have more of them than other berry fruits, and it's that unique mix that seems to be especially healthful. The other thing about wild blueberries is their small size: the pigment is more concentrated in their skin, so cup for cup, you get more anthocyanins from the little blueberries than the big ones." Increased levels of antioxidants in the blood, Camire explains, help reduce inflammation and damage to blood vessels and DNA.

Klimis-Zacas, Camire's colleague, made news in 2010 when she found that blood vessels in the arteries of rats became more flexible when the animals were put on a blueberry-enriched diet. "The effect is similar to blood pressure medication," Camire says. "Stiffening of the arteries is a problem with high blood pressure. If they're flexible, they let blood flow more easily. The next step is to try the experiment with human beings who have elevated blood pressure: if we feed them blueberries, can we bring their blood pressure down?"

Blueberries may also ward off diabetes by improving blood sugar levels and the way the body uses insulin. "We have a lot of animal research to support that finding," Camire says. "We gave rats freeze-dried blueberries or blueberry extract. Anthocyanins have been shown to block the enzyme that digests starch, so the starch is digested more slowly, resulting in a more gradual release of blood sugar. That would be very beneficial to people with diabetes. The difficulty is in extrapolating from rats to human beings and figuring out what the dose and the timing should be: Do you have it with a meal? Do you have it before? Should you have a shot of blueberry juice, and then eat your cornflakes?"

How many blueberries do you have to eat to benefit from their healthful properties? Camire, who prefers whole fresh or frozen berries over juice or extract capsules, suggests one or two cups a day. "I'm a whole foods person," she says. "The juice is good, but it has no fiber. Blueberries have soluble fiber, which in itself is good for lowering cholesterol and blood glucose levels. An extract might have more potency for specific health effect, but the whole fruit offers the most benefits for overall health. Whole foods also give you pleasure, and that's a health benefit, too."

Blueberry Breakfast Smoothie

Soy milk is rich in isoflavones, which reduce cholesterol and the risk of prostate and breast cancers. Like blueberries, whey protein may have anti-inflammatory properties; it has a slightly sweet flavor.

1 cup wild blueberries, fresh or frozen
¾ cup vanilla soy milk
1-2 tablespoons honey
⅓ cup whey protein powder
Dash of fresh ground nutmeg or cinnamon

In blender, combine soy milk and honey. Add blueberries and whey protein and puree until smooth. Season with a dash of cinnamon or nutmeg. Serves 2.

The Rat Olympics

Long dismissed as nutritional wimps, blueberries' rise to the top of the superfoods list can be traced to the relaxing of dietary supplement laws in the mid-1990s. Bilberry extract capsules have been sold in Europe as an over-the-counter drug for improved vision and eye health for decades, so their arrival on health-food store shelves in this country inspired a few researchers to take a look at the bilberry's close cousin, the lowbush blueberry (ripe blueberries have pale green flesh, whereas bilberries are purple throughout, a sign of their higher concentration of anthocyanins). Among them was Jim Joseph, a neuroscientist conducting brain research at the Jean Mayer USDA Human Nutrition Research Center on Aging at Tufts University in Boston.

Excited by a colleague's research into the antioxidant strength of various foods, Joseph began grinding up high scorers like strawberries, spinach, broccoli, and blueberries, and feeding them to aging rats (they were in the human equivalent of their seventies). The animals, he found, performed significantly better on short-term memory tests.

"We also devised a series of motor-coordination tests that we called the Rat Olympics," Joseph wrote in *The Color Code: A Revolutionary Eating Plan for Optimum Health*, which he co-authored with Dan Nadeau, an endocrinologist who at the time was clinical director of the Diabetes Center and Nutrition Support at Eastern Maine Medical Center, and Anne Underwood, *Newsweek*'s health and medicine reporter. "In these tests, we made rats walk miniature planks and balance on slow spinning rods. Amazingly, blueberries were actually able to reverse motor deficits in these aging animals! Blueberry-fed plank walkers retained their balance for eleven seconds — versus just six seconds for those on the standard diet. The results of the "lumberjack test" were even more impressive. On average, the blueberry-fed rats were able to stay on for nine seconds — more than twice as long as the control group on the standard diet."

Joseph's discoveries were the inspiration for many of the studies that followed. Just before his death in 2010, he published a study that showed the polyphenols in blueberries boost brain function by acting as an anti-inflammatory. His influence endures in the annual Wild Blueberry Summit, which he helped found.

The Potent Quotient

The Japanese have become enthusiastic consumers of wild blueberries since their healthy antioxidant properties were discovered about fifteen years ago. They wolf down blueberry pasta, blueberry pizza, and blueberry curry, along with many of the same products westerners enjoy.

"Blueberry jam is the second most popular jam in Japan," reveals Ed Flanagan, CEO and president of Jasper Wyman & Son, the world's second largest grower and processor of wild blueberries. "Blueberry yogurt is also very popular."

So, too, is blueberry wine, as UMO food scientist Mary Ellen Camire discovered. The professor received three complimentary bottles of the stuff from a New Jersey vintner who was convinced that her testimony in the Japanese edition of *Men's Health* magazine had triggered a sales spike in the East Asian nation. What did Camire say that proved so potent? "Blueberries are one of the best foods for older men with erectile problems," she was quoted in the article, titled "The Sex for Life Diet."

No one has directly studied blueberries' effect on erections ("my assistant said that's when he would quit," Camire says, laughing), but the professor bases her conclusions on scientific research. "Viagra works by opening up the blood vessels, and we know from our work and the work of other labs that that's exactly what blueberries do. If something opens up the blood flow in your fingers, it's going to open it up in your other extremities."

Blossoms, Bees & Berries

A career in blueberries was inevitable for Cary Nash. He learned to stand in a playpen in a blueberry field while his mother managed a harvesting crew for her dad, a blueberry grower and processor in the small upcountry town of Appleton, Maine. By age eleven, Nash was working on the crew himself.

Today Nash is what's known as a "large small blueberry grower." He husbands eight hundred acres in fifty fields spread among fourteen midcoast Maine towns, including some of the same Appleton acreage his grandfather managed. (By contrast, large growers like Cherryfield Foods and Jasper Wyman & Son have several thousand acres.)

Like most commercial growers, Nash prunes his blueberry shrubs by burning or mowing them in a two-season rotation: half of the fields are bearing berries to be harvested while the other half recovers from a recent pruning.

In addition to growing blueberries, Nash owns an Appleton receiving station, where his harvest and those of other growers are weighed and shipped to Cherryfield Foods' processing plants in Cherryfield, Maine, and Oxford, Nova Scotia.

If we could peek over Nash's shoulder, his monthly planner would look something like this:

NO.

DATE

January

Offer testimony on bills relating to farming before state legislative committees in Augusta. Go to the Maine Agricultural Trade Show, also in Augusta, to view the latest farm equipment, hear presentations on growing, harvesting and processing techniques, and exchange ideas and news with other farmers.

February

Send letters to landowners reaffirming the grower-landowner partnership. Repair and tune trucks, tractors, and other equipment. With notes from last season in hand, walk the fields and assess what needs to be done.

NO.

DATE

March

Continue appraising fields. Order fertilizers, herbicides, pesticides, and fungicides. (These are generally applied to fallow fields, meaning crop-bearing fields are a year or more past the last application.)

April

Resume burning and mowing begun last fall. Scout for damaging insects and weeds in fallow field; spray pre-emergent herbicides, fungicides, and pesticides if warranted. Set up sticky yellow boards baited with sweet-smelling ammonia acetate to trap blueberry flies, which lay their eggs on blueberries.

A major pest, blueberry maggots tunnel into the fruit and liquefy its flesh. Although the traps do kill the adult flies, growers use them primarily to monitor infestation so they know whether to treat the field with insecticide.

NO.

DATE

May

"Put on bees." Beekeepers bring in three hives to work the four hundred fruit-bearing acres for three to five weeks. Scout for mummyberry, a fungus that attacks flowers and leaves. Make note of any found for later treatment; it is too close to harvest to apply fungicide. Continue spraying fallow fields.

June–mid-July

Spread sulfur and fertilize fallow fields.

July

Deliver portable toilets and blueberry crates to the fruit-bearing fields for

Regulars include a Louisiana labor contractor and his crew of twenty-five Mexican migrant workers, who camp at Toddy Pond in Oakland. Another fifteen or twenty rakers are hired locally. Rounding out the crew are six workers with harvesting tractors and five or six truck drivers who will deliver the rakers' crates – each one can hold about twenty-five pounds of berries – to the Appleton receiving station.

DATE

the harvesting crews who arrive late in the month.

The tractors are here for only a week, after which they move on to the blueberry barrens of Washington County.

August

Begin harvesting, if it hasn't already started. Dispatch harvesting machine operators to fields that are level and cleared of rocks. Send hand rakers to work the rocky, steep fields, which in this part of Maine means most of them. One hundred thousand pounds of blueberries – some from other area growers – will come into the receiving station on the first day. There, an employee weighs them – workers are paid $2.25 to $2.50 a

NO.

DATE

pound – and stacks the crates on the loading dock. Load the day's bounty into fifty-three-foot-long trucks bound for Cherryfield Foods' two factories. Within twenty-four hours of picking, most of the blueberries will be frozen.

When the harvest is done at the middle of the month, collect any crates left in the fields, "wipe" weeds (that is, apply herbicide), and have the portable toilets removed.

September

Wipe weeds. Bush hog and mow some of this season's fruit-bearing fields;

NO.

DATE

they are next year's fallow fields.

October

Mow. Burn fields that can't be mowed.

November and December

Mow if weather permits.

The fields are set afire with a blower oil burner that resembles a simple cannon hitched to a tractor. It roars like a locomotive as it throws a five-foot flame. A worker trails behind with a water tank, dousing stray flames. The work is grimy ("You have black boogers for days," Nash says) and expensive, about $400 per acre depending on the price of oil.

*

From *Cooking Down East* by Marjorie Standish

2 cups blueberries
juice of half a lemon
½ teaspoon cinnamon

Preheat oven to 375°F. Butter an 8-by-8-inch pan; turn blueberries into pan, dribble lemon juice over them, and sprinkle cinnamon over berries.

¾ cup sugar
3 tablespoons margarine
1 cup sifted flour
1 teaspoon baking powder
¼ teaspoon salt
½ cup milk

Cream margarine and sugar and add sifted dry ingredients alternately with milk. (There are no eggs in the recipe.) Spread the batter all over the top of the berries.

1 cup sugar
1 tablespoon cornstarch
dash of salt
1 cup boiling water

Mix sugar, salt, and cornstarch. Turn this dry mixture all over the batter. Then pour 1 cup of boiling water over the top.

Bake for 1 hour. Serve warm, topped with a serving of vanilla ice cream or whipped cream. It is also especially nice served plain. Serves 8.

TOWER

Man vs. Weeds: The Organic Blueberry Farm

Organic blueberry farmers follow the same schedule that growers like Cary Nash do, but their spring is more labor intensive. "Our goal is to stay on top of the weeds," says Hugh Lassen, who with his wife, Jenny, tends seven-acre Intervale Farm in Cherryfield. The Lassens are founding members of the Down East Organic Blueberry Alliance, a cooperative of organic growers working together to increase availablity of their produce. "We weed with a scythe or a weed wacker in May, June, and July. We have to watch for diseases because we can't spray for that and an outbreak could devastate the crop. That's one of our big challenges."

Neither Hugh or Jenny come from farming backgrounds. They were both working for a parks organization in New York City — she as a gardener, he as a carpenter — when they decided to head northeast in search of land and a new way of life. "We were looking for something small, a farm on a family scale," says Hugh, "and we were thinking fruit — strawberries, maybe, or raspberries. We found this blueberry field in 2005 just after the harvest. We felt that as a native crop it would be easier than trying to bring something in."

The land had been in blueberries for more than fifty years, most recently managed by giant grower and processor Cherryfield Foods. Because the field had been sprayed with

Intervale Farm, an organic blueberry farm in Cherryfield.

pesticides, the Lassens had to manage it without any of those products for three years before it could be certified as an organic farm. They sought advice from the Maine Organic Farmers and Growers Association (MOFGA), University of Maine's Blueberry Hill Farm, and their blueberry-growing neighbors.

Like most growers of lowbush blueberries, the Lassens harvest only half the field each year, netting an impressive ten thousand pounds of berries from 3½ acres every August. They mow to prune, but burn every few years by spreading the fields with straw and igniting it. "Burning periodically is recommended to kill diseases and bugs," Hugh explains.

The Lassens' raking crew typically includes a few friends, Jenny's parents, MOFGA apprentices, and volunteers from World Wide Opportunities on Organic Farms, a worldwide network that links organic farmers with travelers who work in exchange for room and board. Hugh winnows the berries the day they are picked with a small machine that was built by Rockland inventor Emil Rivers in the 1930s. Originally cranked by hand, the winnower was outfitted with a motor some time during its long life.

Most of the Lassens' berries are sold fresh and frozen through the alliance, and Jenny makes fruit spreads and chutney that she sells under the label Tin Penny Jenny from a farmstand and at farmers' markets. "We're still learning and figuring things out as we go," Hugh says. "It gets easier every year."

Organic blueberries, jam, and chutney from Intervale Farm.

From *Superb Maine Soups* by Cynthia Finnemore Simonds, greet your morning full of energy and antioxidants with this breakfast soup. Wyman's Wild Blueberry Juice—available at most stores—is 100% blueberry juice. Serves 4.

- 1 cup wild blueberries, fresh or frozen
- 2 cups organic vanilla yogurt
- 3 bananas
- 2 cups blueberry juice
- 2 T golden flax seeds
- ¼ cup wheat germ
- banana slices and whole blueberries for garnish

Put all ingredients, except the garnish, in a food processor or blender and pulse until smooth. Serve garnished with the banana slices and whole blueberries.

Intervale Farm

What's the Buzz?

Wander into a wild blueberry field in May, when the new green foliage is dusted with bell-shaped white flowers, and you will hear it: a low, steady hum, as if a jet were warming up just beyond the farthest hill. You soon realize, however, that the soft buzz is not far off at all. It's all around you. You can even see it, seemingly bending the air over the blossoms, like heat rising from pavement.

Bees. Thousands upon thousands upon thousands of bees.

The unsung heroes of the commercial wild blueberry crop, honeybees are a large part of the reason Maine's harvest has quadrupled over the past twenty-five years. Growers no longer rely solely on native bees to pollinate their blossoms. Instead they hire the services of migratory commercial beekeepers, who deliver 65,000 hives — that's nearly four *billion* bees — to Maine's blueberry fields and barrens each spring. The only crop that demands more of the little workers is California almonds, and few require more of their time — wild blueberries have an unusually long pollination season, about three to four weeks, because every field contains dozens of clones, each blossoming according to its own timetable.

Once the blueberries are pollinated, most of the beekeepers will move on to another crop in another state — cranberries on Cape Cod, maybe, or alfalfa and lima beans in the Midwest — but a few, like Lincoln Sennett, stay behind. Some of Sennett's eight hundred hives will be trucked directly to his Swan's Honey processing plant in Albion, where the blueberry honey will be extracted, filtered, and bottled. The other hives will be sent to perform their busy dances elsewhere, perhaps wild raspberry blossoms in Aroostook County, after which they, too, will be trucked to Albion and relieved of the luscious, golden syrup they make from nectar.

As a honey producer and blueberry grower, Sennett is unusual in the world of modern beekeeping. "Most commercial beekeepers in the United States are not in the honey business," he says. "They make their living off services like pollination. Some of the bigger guys have eight to ten thousand hives. When you get that big, you become a truck driver. Half your business is moving the bees."

Sennett prefers the challenge of managing the only insects that make a food that is consumed by humans. The practice of forcing bees from one monoculture environment to another, he adds, has been implicated in colony collapse disorder, the catastrophic loss of honeybees worldwide. "Bees need a variety of pollens to be healthy," he believes. "They need a balanced diet because many

Lincoln Sennett holds a frame from a honeybee hive

pollens are deficient in the amino acids that they need. It's just like us: We wouldn't be healthy if we ate just one thing, like bread."

So it is that in winter and early spring, when his competitors' bees are working in California's almond orchards or hopping up the coast from North Carolina cucumbers to New Jersey high bush to New York apple orchards, Sennett is building up his bee populations in South Georgia and monitoring a few hibernating hives in Maine, where he leads workshops for home beekeepers. When he does move his hives north in May, Sennett sends them not to the vast barrens but to forty or so small blueberry fields, including his own 150 acres in Machias. "They're flying into the woods and bringing back all different kinds of pollens so they're less likely to have the amino acid deficiencies," he believes.

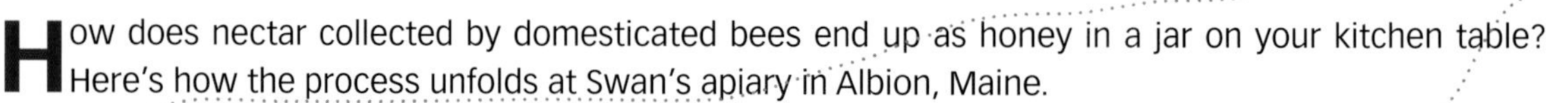

Follow the Honey

How does nectar collected by domesticated bees end up as honey in a jar on your kitchen table? Here's how the process unfolds at Swan's apiary in Albion, Maine.

1. The honeybee sucks nectar from the blueberry blossom into her abdominal honey sack and flies it to the box hive, where she is met by her co-workers. They transfer the nectar into honeycomb cells on the hive's frames.

2. Another group of worker bees have the job of fanning their wings over the combs so water in the nectar evaporates, creating pure raw honey. The bees then seal the comb cells with beeswax. When 90 percent of the frame is capped, it is ready to be harvested. A frame contains about four pounds of honey; an entire hive, about forty pounds.

3. When most of the comb cells are capped, the beekeeper places a fume pad on top of the hive. Its stinky odor drives the bees down to the bottom of the box, so the apiarist can remove frames without getting stung. With a machine called a flail uncapper, the apiarist removes the wax cappings. They'll be melted down and used for lip balm and other products.

Blueberries may help reduce belly fat, according to a study by the University of Michigan Cardiovascular Center. Rats fed a blueberry-enriched diet have less abdominal fat, lower triglycerides, and lower cholesterol.

A diet rich in blueberries may improve motor skills and reverse the short-term memory loss that comes with aging and age-related diseases such as Alzheimer's, according to studies conducted by the USDA Human Nutrition Research Center on Aging.

4. The frame is placed in a centrifuge, which spins the raw honey out of the wax cells and into a pipe that drains into a tank. Stored in drums, the honey contains pollen grains and wax particles that will cause it to crystallize quickly. If the combs are in good condition, they'll be returned intact to the hives so the bees can go straight to making honey, rather than rebuilding their home.

5. The drums are inverted and heated to liquefy the honey. It spills into a stainless steel tank, from which it is pumped into the packing room. There the honey passes through a filtering system to remove pollen, wax particles, and any other debris that can cause crystallization.

6. The honey flows into a bottling tank, and, finally, into individual bottles.

Duck Breast with Wild Blueberry Sauce

From *Wild Blueberry Association of North America*

- 1-inch piece fresh ginger, diced
- ¼ cup Mirin
- ¼ cup soy sauce
- 2 tablespoons granulated sugar
- cayenne pepper
- 2 duck breasts
- 1 tablespoon vegetable oil
- 2½ cups wild blueberries
- ½ cup duck stock (or chicken stock)
- 2 teaspoons cornstarch
- salt to taste
- 8 ounces asparagus, trimmed and steamed

In small saucepan, bring Mirin, soy sauce, and sugar to simmer and cook for 5 minutes. Add cayenne pepper to taste; let cool. Place duck breasts in shallow dish and pour Mirin mixture over and turn to coat. Cover with plastic wrap and refrigerate for 2 hours.

In skillet, heat oil over high heat and brown both sides of breasts. Place on a baking sheet, skin side up, and roast in 400°F oven for 15 minutes. Let cool 5 minutes, before slicing.

Return skillet to medium-high heat and add blueberries. Add remaining marinade and duck stock. Cook, stirring until boiling. In small bowl, whisk cornstarch with 1 tablespoon of cold water and stir into sauce. Cook, stirring for 1 minute or until thickened. Add salt to taste. Serve sauce with sliced duck breast and asparagus. Preparation time: 15 minutes plus 2 hours marinating. Cook time: 20 minutes. Serves 4.

Grunts of Satisfaction

We have colonial cooks to thank for these comforting blueberry desserts, all variations on pie.

Buckle: A two- or three-layer baked dessert. The bottom layer is cake batter mixed with blueberries or topped with a layer of the fruit. The streusel topping is made from flour, butter, and sugar.

Cobbler: A baked deep-dish dessert with blueberry filling and biscuit dough topping. The dough is either dropped onto the filling in spoonfuls or spread evenly over the fruit.

Crisp: A cobbler with crumb topping. The topping is most often made with flour, oatmeal, brown sugar, nuts, and butter.

Grunt: Blueberries steam-cooked on the stovetop beneath drops of biscuit or dumpling dough.

Slump: See Grunt. (Some cookbooks say a slump is a baked grunt, but doesn't that make it a cobbler?)

Jennifer Smith-Mayo

Wild Blueberry Country

One of the kitschiest landmarks on Route 1 in Maine is a giant, grin-provoking blue dome, part amusement park, part specialty shop, bubbling up from the earth in the Down East town of Columbia Falls. Laying claim to the title of "world's largest blueberry" (though one might argue it better resembles a flying saucer for Smurfs), Wild Blueberry Land sits, appropriately enough, in the heart of the most concentrated collection of blueberry barrens in Maine. It is the playful invention of Marie and Del Emerson, who are as steeped in wild blueberry culture as any Washington County family gets. "I didn't go to college with blueberries in mind, but in this neck of the woods, there were only three things you could do: forestry, fishing, or blueberries," says Del, who was born in a house without electricity on the banks of Columbia Fall's Pleasant River in 1935.

Del Emerson is a big name in Maine blueberries. He spent fifty-six years at the University of Maine's Blueberry Hill research farm in neighboring Jonesboro, thirty-five of them as manager, and he is the inventor, with son, Zane, of the Emerson Harvester, a hand-pushed blueberry picker that can collect twenty-five pounds of berries in just thirty seconds.

Marie occupies a rather different niche in wild blueberry culture. The American

Culinary Federation's first woman Chef of the Year and Maine's only certified research chef (an occupation that blends culinary arts with product research and development), she is a culinary and baking instructor at Washington County Community College in Calais. She and her students continually develop new products using Maine-grown foods like potatoes, salmon, and, of course, wild blueberries, which have found their way into recipes like rangoons (deep fried Asian-style dumplings with cream cheese filling) and chilled soups. Not surprisingly, Marie is regularly called upon to judge food entries at the Machias Wild Blueberry Festival (the most unusual entry she ever sampled: baked lobster with blueberry and crabmeat stuffing, page 49). All that, and she is the queen of Wild Blueberry Land, too.

The bakery is the main attraction at Wild Blueberry Land, filling the shop with the aroma of cookies, muffins, scones, and pies. "Every pie is different because each one is homemade one at a time," Marie says proudly. "The scones are fresh. The muffins are fresh. We do everything by hand, no machines." The shop carries every sort of blueberry product imaginable — jellies, honey, cookbooks, souvenirs — and Marie is always scouting for more. "We built this place to add value to other people's products," she says.

Reminiscent of 1950s roadside architecture, Wild Blueberry Land is in fact a relatively new landmark, built in 2000. "This corner has been a motel, a bar, and a permanent yard sale," says Marie, who is wearing a blueberry-stained apron and sitting at a picnic table outside the big blue dome. "The lot kept changing hands, so I finally told Del, 'We need to buy that corner. I want to put up a giant blueberry.'"

They tracked down a manufacturer of geodesic domes and built the blue monster themselves, spraying the plywood exterior with foam and sealing it with silicone. "We painted it blue," Marie recalls, "and within twenty-four hours, the paint had peeled off. Turns out, you can't paint silicone. Nothing will stick to it." Undeterred, they found a dyed silicone, which looked great at first, but faded within a year. The third try, with a dyed UV silicone, was the charm. "It'll never leak," says Marie, happily pointing out the silver lining in their painting challenge. "It's got three layers of silicone!"

Del Emerson and his Emerson Harvester

The Emersons have continued to build their blueberry world, scattering several large round buoys (painted blue, of course) around the parking lot and opening a small blueberry-themed miniature golf course. Marie hopes eventually to move Del's winnowing and packing facility to the site. "This is my Disney World," she says wryly, "This is a fantasy. It's flowers and it's beautiful and it's play."

From *Good Maine Food* by Marjorie Mosser

- 1 quart blueberries
- 1 cup sugar
- ¼ cup melted butter
- juice of half a lemon
- 1 cup cake flour
- 3 teaspoons baking powder
- ¼ teaspoon salt
- ¼ teaspoon salt
- pinch of nutmeg
- 3 tablespoons lard
- 1 egg
- ¼ cup milk

Put blueberries, sugar, butter, and lemon juice in a baking pan. Mix and sift dry ingredients and cut in lard. Beat eggs and milk together, then stir into dry ingredients. Cover the blueberries with this mixture, and bake in a moderate oven, 350°F for 40 minutes. Cut in squares and serve with whipped cream or ice cream.

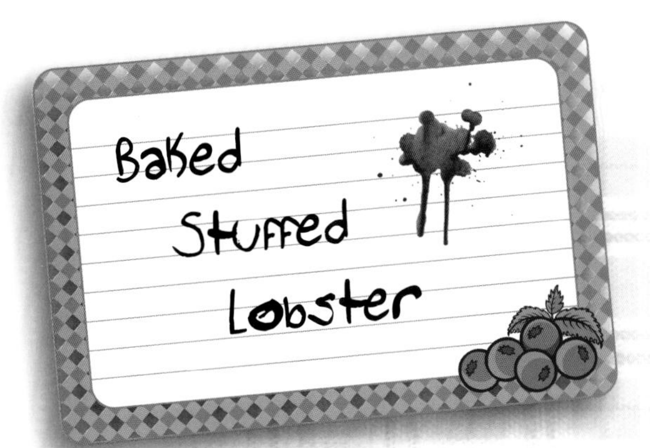

Blueberries and lobster are quintessentially Maine, but who'd think of combining them in one recipe? Catherine Ryan Quint did, and her creation was a prize winner at the Machias Wild Blueberry Festival.

Two 1½-pound lobsters, ready for stuffing
4 tablespoons butter or margarine
1 clove garlic
¼ cup chopped onion
2 tablespoons dry white wine
1 cup blueberries
2 cups fresh bread crumbs
salt and pepper to taste
2 tablespoons chopped fresh parsley
½ cup crabmeat (optional)

Melt butter; add garlic and onion and saute until soft. Stir in parsley and wine; cook 2 minutes more. Add bread crumbs, crabmeat, and blueberries. Remove from heat. Season to taste. Stuff body cavity of each lobster and place in shallow baking pan. Bake in preheated 400-degree oven for 20-30 minutes. Serves two.

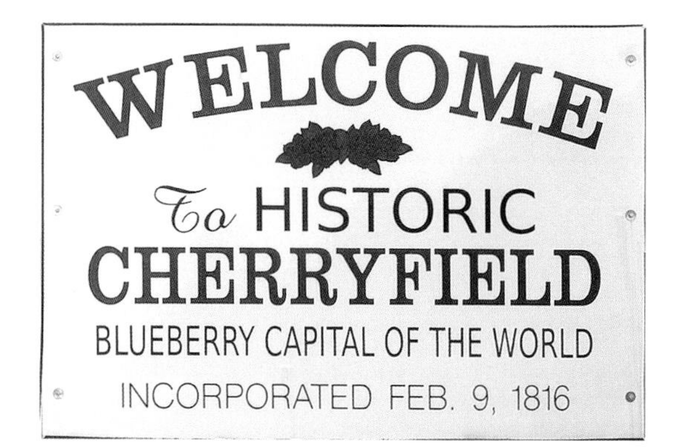

The Blueberry Capital of the World

The self-proclaimed Blueberry Capital of the World is ... Cherryfield? It's true. The Down East town is home to the factories of the world's largest blueberry growers and processors — Jasper Wyman & Sons, a Maine company, and Cherryfield Foods, a Canadian company owned by Oxford Frozen Foods — and the gateway to the state's largest stretches of barrens, sprawling northeast into Deblois, Columbia Falls, Epping, Centerville, and Jonesboro.

Trees, not blueberries, attracted Cherryfield's early settlers. "There were acres and acres of tall straight white pine, huge expanses of forest that nobody had touched," Kathy Upton, president of the Cherryfield-Narraguagus Historical Society, says. "Lumber was king."

Harvested logs were floated on the Narraguagus River, which flows through the picturesque village, where they were sluiced through a series of nine dams, each one supporting a couple of mills. "This was a booming community," Upton says. "There were nearly two thousand people here in the late 1800s." Today, 1,157 people call Cherryfield home.

As lumbering faded in the late nineteenth century, blueberry farming picked up, aided by the timber harvesting that had opened up barrens on the north side of town. The lumber barons' legacy is the thirty or so grand houses, some quite flamboyant, that line the leafy banks of the Narraguagus, which is quiet these days except for the springtime aerial dances of bald eagles and osprey diving for alewives and salmon.

As for the town's name, it comes from the wild cherries that grow along the river's banks. "They are different than bunch cherries," Upton says. "They are single cherries on a stem and they're bright red. They make the most delightful jelly if you have the patience to pick them."

Capturing a pint of blueberries at Del Emerson's fresh pack facility in Addison.

Wild Blueberries with Roquefort, Celery, and Cucumber

From *Wild Blueberry Association of North America*

- 2 cups of frozen wild blueberries
- ½ cup walnuts
- ½ cup Roquefort or blue cheese, crumbled
- 1 bunch of celery
- juice and zest of half orange and half lemon
- ⅓ cup of ruby port
- 1 teaspoon currant jelly (or other fruit jelly)
- pinch of cayenne

Cumberland Sauce: Combine citrus ingredients, port, jelly, and cayenne in small sauce-pot and simmer until reduced by half or until thickened enough to coat a spoon.

Salad: Let blueberries defrost. Chop walnuts into small pieces and roast them gently in a coated pan without fat. Divide the Roquefort into bits. Clean and wash celery. Cut into 3-inch pieces. Fill celery pieces with combined mixture of blueberries, walnuts, and cheese. Drizzle each filled celery piece with Cumberland Sauce.

Alternatively: Cut pieces of celery into 1/2-inch pieces, toss lightly with cheese, walnuts, and blueberries. Drizzle with Cumberland sauce and serve as a salad on a leaf of Boston lettuce.

From *Wild Blueberry Association of North America*

½ cup frozen wild blueberries
4 ounces white rum, optional
3 tablespoons sugar
10 ounces soda water, very cold
2 limes, juiced
3 ounces coconut cream, warmed to soften
presentation: 1 egg white
presentation: fine table salt

To prepare the cocktail glasses, dip the rims in egg white, then in fine table salt. Set aside. Place the blueberries in a blender with the rum (if using), sugar, soda water, half of the coconut cream, and the juice of two limes. Pour into prepared cocktail glasses, then finish with a swirl of remaining coconut cream. Makes 4 cocktails.

Jennifer Smith-Mayo

Celebrating Blueberries

A Wild Blueberry Queen has no time to rest on her laurels. "Right after the coronation she goes off to attend various events at the fair," says pageant organizer Yvonne Drown. "She introduces activities and serves diners at the pancake breakfast. She is present at all the cooking contests, where she gives out the awards."

The queen's coronation is a cornerstone of the Maine Wild Blueberry Festival, an eight-day event that begins on the third Saturday in August. Part of the Union Fair in the midcoast town of Union since 1959, it is Maine's oldest blueberry celebration, but not the largest. That honor belongs to the Machias Wild Blueberry Festival, which draws 15,000 people to the Down East town of Machias on the third weekend of August. There are many summer fairs that give blueberries a nod with pie-baking and pie-eating contests, but only Machias and Union make them the cause célèbre.

Typically about ten young women — "princesses" between the ages of seventeen and twenty-two — compete for the title of Wild Blueberry Queen, which comes with a $1,500 scholarship, a trophy, roses, and a slew of duties for the coming year. "It's not a beauty pageant, though beauty and poise do play a role," Drown says. "Contestants appear before a panel of judges to give a presentation

The Children's Parade opens the Machias Wild Blueberry Festival.
Opposite page: Children wolf down pie at the Machias Blueberry Festival.

about their life's story and to tell how they feel about representing the blueberry industry."

Strong public speaking skills are important, and no wonder, given the queen's responsibilities. In the weeks immediately following the fair, she is expected to deliver a presentation to the Wild Blueberry Commission of Maine and to represent the state at the Big E, or the Eastern States Exposition, in West Springfield, Massachusetts. Her one-year reign continues with appearances at fairs, parades, and charitable events, tours of blueberry processing facilities, and meetings with state legislators.

Drown was a princess herself in 1971, and her daughter, Janelle Drown Thompson, was selected Crown Princess (first runner-up) and Miss Congeniality in 2001. Janelle also has served on the judge's panel.

Mother and daughter come to their roles with good credentials: Blueberry Rock Farm in Hope has been in the Drown family for 150 years. One of dozens of small farms dotting the hillsides of inland Knox County, Blueberry Rock produces sixty tons of blueberries every other year.

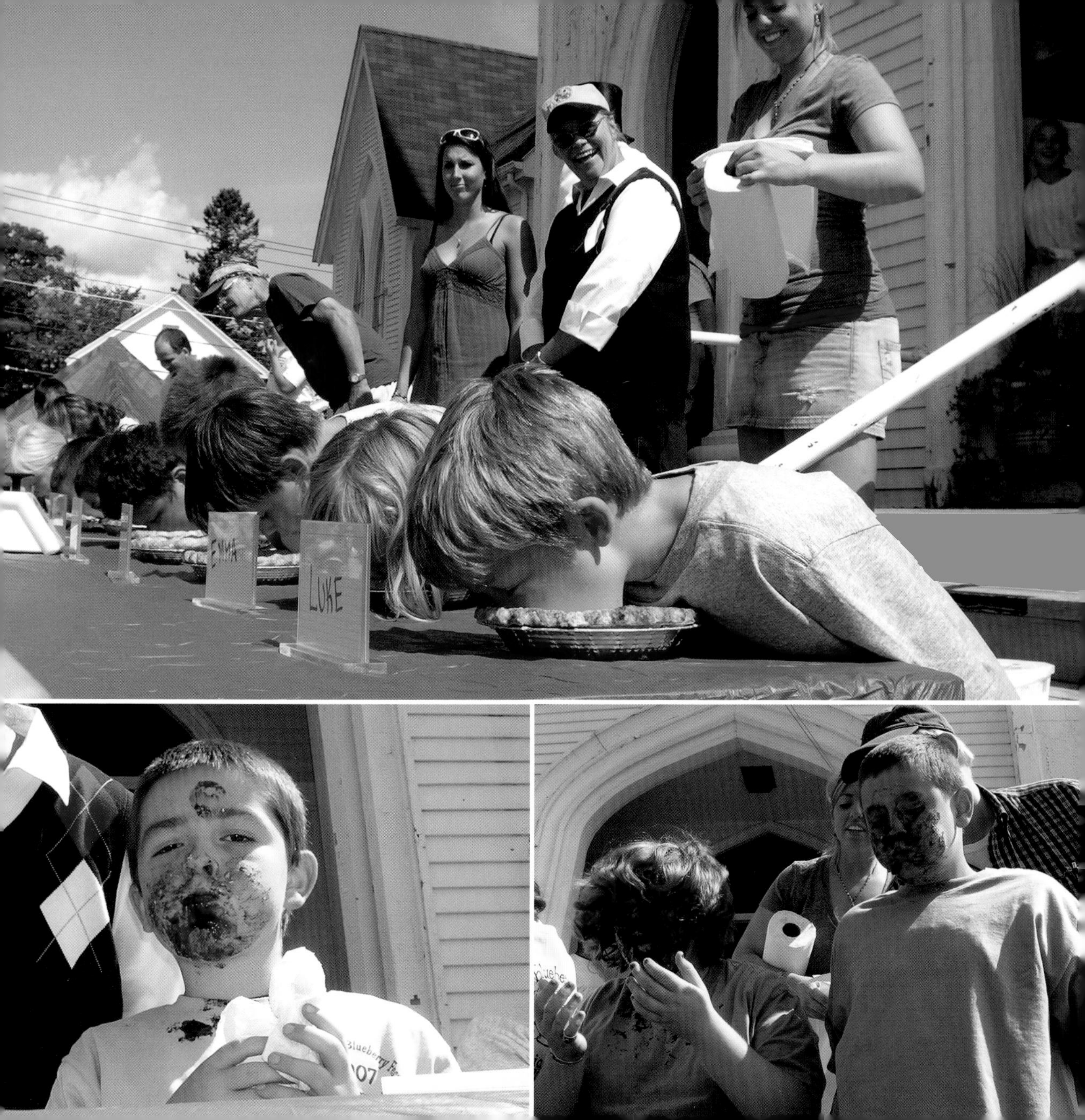
EMMA
LUKE

Blueberry Queen Bethany Snow

Three months after she graduated from Thomaston's Georges Valley High School in 2010, Bethany Snow was crowned Maine Wild Blueberry Queen at the Union Fair's Wild Blueberry Festival. She is a student at Salve Regina University in Newport, Rhode Island, and is contemplating a career in health care. As the Wild Blueberry Queen, her duties are to represent the Maine blueberry industry at festivals and special events for one year.

Why did you want to be the Wild Blueberry Queen?
Growing blueberries is such a big part of Maine culture. I also was interested in the $1,500 scholarship.

How did you prepare for the pageant?
The pageant organizers sent a packet of information about blueberries, and I studied it. I also had to write an essay and present it at a picnic attended by my family, my friends, and my pageant sponsors, the owners of Scott's Blueberry Hill Farm in Waldoboro.

Was the essay about blueberries?
No, it was a personal essay about my experiences and some of the milestones in my life, including graduation and sports — I've been involved in two softball state championships and one soccer state championship.

Describe meeting the judges at the Union Fair.
It took place in the Blueberry Acres building, and there were three judges and a small crowd watching. Each of the ten contestants had to stand up and the judges asked us about blueberries and Maine and about our aspirations and goals.

Were you nervous?
A little. But once you start talking you get used to it. I'd done a little public speaking in high school, but never in front of as many people as this.

Did you have a sense that you'd won?
Not really. I felt like it was up in the air, that it could have been any one of us.

Have you ever raked blueberries?
I raked for the first time after I entered this pageant. My sponsors invited me to spend a day with them. I raked out in the field for half the day and worked the picking line, picking out bad berries, for half the day. Both are hard work — you wouldn't think the picking line would be as hard as raking, but you are bent over much of the time and you have to be quick.

Do you like blueberries?
I love blueberries! You couldn't be the Blueberry Queen if you didn't!

Blueberries were first harvested commercially in the 1840s. During the Civil War, sardine canneries, deprived of their southern market, began processing vitamin C-rich blueberries, which were shipped to Union soldiers to prevent scurvy. The demand for blueberries rose when those soldiers came home.

Singing the Blues

He must be from away," the residents of Rakealot rightly conclude when Lancelot comes trotting into town. If his chain-mail tunic and his horse named Taxi hadn't already given him away as an outsider, his grasp of blueberry matters soon would have. "I understand you rake blueberries here," the knight from New York says with just a hint of patronization. "That's new to me. Where I come from, we have *highbush* blueberries."

At this, mocking groans ripple through the packed pews of the Centre Street Congregational Church in the very real wild blueberry country hamlet of Machias. Having been warmed up by the saddle shoe-wearing, ukulele-playing conductor Gene Nichols and his fifteen-piece band, the audience is in a merrymaking mood as the blueberry musical comedy, the signature event of the annual Machias Wild Blueberry Festival, gets under way.

Created in 1975 by the Centre Street Congregational Church, the Machias festival has grown to involve the entire town. Restaurant marquees trumpet blueberry pancakes and blueberry pies, shops sell all manner of blueberry souvenirs, and motels book solid a year in advance. The opening night fish fry dinner is hugely popular, as are the blueberry

Don't rinse your just-picked wild blueberries until you are ready to eat or use them because moisture shortens their storage life. Do, however, pick through the berries and remove any that are broken or smashed. They will stay fresh in an air-tight container in the refrigerator for about a week.

Freezing wild blueberries is easy. Do not rinse them as this will cause them to stick together in a frozen clump. Pick through the berries and remove any that are broken or smashed, then put the berries into re-sealable plastic freezer bags and freeze. Rinse the berries when ready to use.

---

pie-eating contests — there are four of them, ten contestants at a time. But it's the musical comedy that truly defines the festival's spirit, not to mention that of the resilient, hardworking people who live in this achingly beautiful, yet very poor, part of Maine.

Consider 2002's *Gone With the Winnower*: Arriving home smeared with purple stains after a long day in the field, lead character Ruby dramatically declared, "As God is my witness, I'll never rake blueberries again!" (The scene spoofed, of course, the iconic moment in *Gone With the Wind* when Scarlet, silhouetted against a brilliant orange sunset, vows, "As God is my witness, I'll never be hungry again!") Then there was *My Blue Hero*, in which the Blueberry Marshall, who wears a velveteen cape and the letters BM on his chest, battles one harvesting crisis after another.

Both shows are among the favorites of their author, Marjorie Ahlin, who has created and directed most of the festival musicals, casting them entirely with local people. "We don't have auditions," Ahlin says. "If you want to be in the show, you're in the show. I try to make sure everyone has at least one line or a solo to sing. We have the best audiences — they are so supportive. The more they laugh and stamp and scream and carry on, the better the cast gets."

So it is with 2010's *Rakealot*, which Ahlin loosely (very loosely) modeled after *Camelot*, complete with altered lyrics from that show's score: "If ever I should leave you," warbles Arthur, who competes with Lancelot for Guinevere's affections. "It wouldn't be in summer. Seeing you in summer I never would go. Your hair full of briars, you fingers so blue, your face so sunburned, and a bite or two...." It's enough to convince Lancelot he will always be "from away" and so that is where he must return. Raising his arm, he calls, "Taxi!" and off he rides.

Dana Moos' Maine Blueberry Malted Belgian Waffles with Maple Syrup and Fresh Whipped Cream

From *The Art of Breakfast* by Dana Moos. There's just something about the flavor of a blueberry waffle served with buttery, pure maple syrup. Simple and delicious!

Makes 12 waffles

2¼ cups flour

2 tablespoons granulated sugar

⅛ teaspoon salt

½ cup canola or vegetable oil

1½ cups 2 percent milk

1 egg

¼ cup malted milk powder

2 tablespoons baking powder

1 cup Maine blueberries (thaw and drain well if you use frozen berries)

Fresh whipped cream for garnish

In a large bowl, combine the flour, sugar, salt, oil, milk, egg, malted milk powder, and baking powder and mix with a whisk until well combined. Do not over mix—it's OK if the batter is slightly lumpy. Gently fold in the blueberries, using caution not to break them.

Heat a waffle iron and liberally coat it with cooking spray.

Ladle the batter onto waffle iron and cook until golden brown—about 8 minutes. To keep the waffles warm and crisp, keep them covered in a 250°F oven and then just before serving uncover to expose waffles to the dry heat for a few minutes, then plate.

Blueberry Blues

Every August brings a brand new blueberry musical to the Machias Wild Blueberry Festival, but one aspect of the show never changes: The finale, in which cast and audience stand and sing "Blueberry Blues," penned by Machias Memorial high-school teacher Norm Dubois back in the festival's early years.

There's a county in Eastern Maine that you can see
One day out of four if you're lucky.
Oh, the fog rolls in with all its might
People and land go out of sight.
Washington County, Maine.

(Chorus)
Blueberry, blueberry, blueberry blues
Blueberry, blueberry, blueberry blues
Oh you rake all day, you rake all night;
You rake, rake, rake 'til they're out of sight.
Washington County, Maine.

Early in August men are out with string
Making rows so we can do our thing.
You take your bucket, you take your pail,
Wait for sunshine, hope it doesn't hail.
Washington County, Maine.

(Repeat Chorus)

Some are fast, some are slow, some just in between
Just get paid for what you do …
You know what I mean.
You take your berries to the machine,
Winnow 'em, winnow 'em 'til they're clean.
Washington County, Maine

(Repeat chorus)

Jennifer Smith-Mayo

Pie Wars

It's 4:30 P.M. on the eve of the last day of the 2010 Union Fair, and a small crowd has gathered around a table laden with blueberry pies. Some in the group are the pie bakers themselves, each one hoping to take home a blue ribbon and a hundred dollars. The pies are beautiful, the crusts a perfect golden brown, edges nicely fluted. Some have fancy touches, like cut-out pastry leaves on top. A few of the crusts are cracked, exposing the rich blue filling underneath. It's an aesthetic flaw, perhaps, but a mouthwatering one.

The chatter softens when the judges stride into the room. Joanne Weatherbee, Amanda Boyington, and Susan Boivin are all wearing aprons. Each carries a clipboard and a bottle of Poland Spring water. There is no introduction; instead, they get right to work, circling the table, looking at the pies, and reading the recipes on the index cards beside them. Boivin, director of the lunch program at Camden Hills Regional High School where she has introduced recipes like chicken masala and Thai peanut shrimp, holds a pie at eye level, turning it. She lifts it higher, peering through the glass dish at the bottom crust. She puts the pie down and marks her scorecard. Her face is impossible to read.

After the judges have had a good look, the contest organizers Irene Maxcy and Lorraine Strout begin to cut the pies, placing

Wild Maine
Blueberries

three slices of each on white plastic plates. The judges slowly trail after them. Weatherbee, a family and consumer science teacher in the Rockland and Thomaston schools, pierces a slice with a white plastic fork and watches the crust flake and crumble. Boyington, who has a home-based bakery in Appleton and manages the fair's Blueberry Hut dessert concession, nudges another slice's filling with her fork, then she tastes it. She follows up with a sip of water and a note on her scorecard. Occasionally, the judges confer quietly. They remain poker-faced. After forty minutes of poking and tasting, they leave to tally points in private. In their wake, the gathering closes in for a closer look at the pies.

Wild blueberry pie contests like this one take place at agricultural fairs all over Maine, courtesy of the Maine Wild Blueberry Commission, which invites the first-place winners to compete at the State of Maine Two-crusted Wild Blueberry Pie Competition in Augusta. The Union Fair also hosts a wild blueberry muffin contest, a wild blueberry dessert contest, and a children's blueberry cookie contest. The Machias Wild Blueberry Festival broadens the challenge with a contest that includes appetizers, breads, pancakes and waffles, punch and

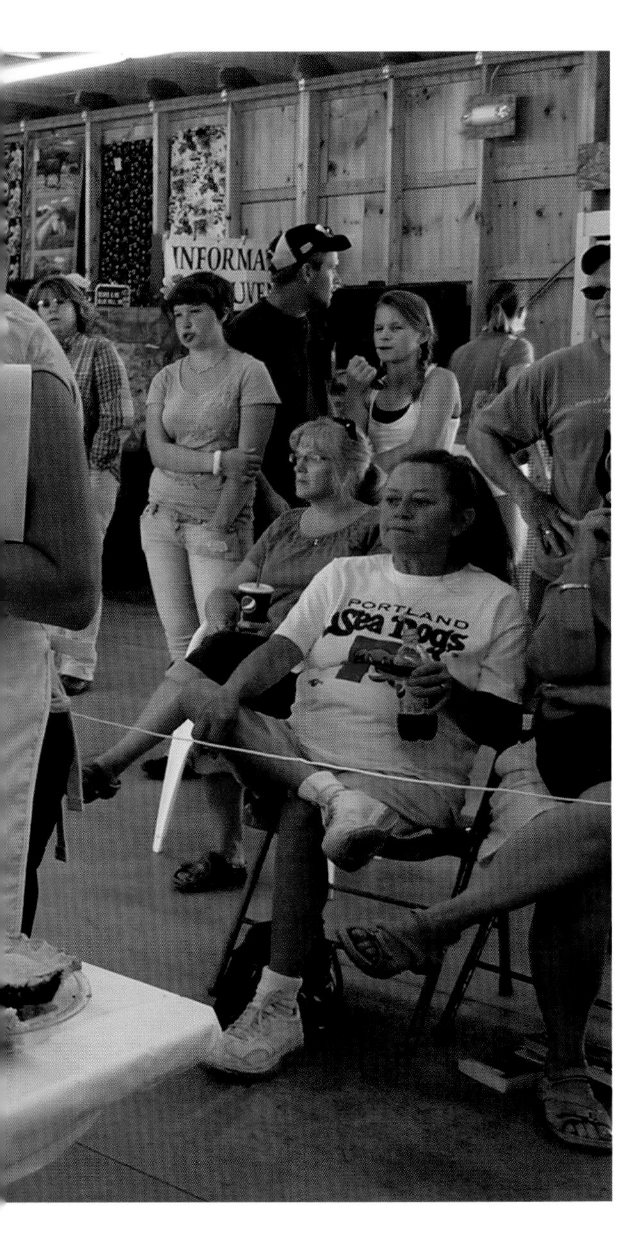

Joanne Weatherbee, Amanda Boyington, and Susan Boivin judge the blueberry pie contest at the 2010 Union Fair.

Pie contest judge Amanda Boyington checks the flakiness of the crust.

wines, jams and jellies, and entrees. The two-crust blueberry pie, though, is the classic, and the rules are these: All pies must be accompanied by the recipe printed on a 4x6 index card; all pies must be submitted in a clear glass pie plate; and the Maine wild blueberry must be the only fruit ingredient. Pies are judged on appearance (thirty points), crust (thirty points), filling (thirty points), and ease and clarity of recipe (ten points).

Back at the Union Fair, the judges have returned. Wild Blueberry Queen Bethany Snow, dressed in a satiny blue dress and a silver crown, welcomes the visitors, whose numbers have grown considerably and include then-Maine First Lady Karen Baldacci. Joanne Weatherbee steps forward to speak for the judges. "It was very difficult to choose," she says. "We base our selection on appearance — eye appeal, uniformity of shape and color of crust — and flavor. We make sure the recipes are intelligible."

It falls to Mrs. Baldacci to announce the winners, beginning with Helen Robinson, who accepts her third-place yellow ribbon and forty dollar prize. Second place winner, Kevin Knapp, the rare male competitor, gets a red ribbon and sixty dollars. Finally, the former first lady calls the name of Faye Harvey of Union, a contest veteran who has collected several blue ribbons over the years, including one for a white chocolate strawberry pie that took first place in 2009 at the Pittston Fair, site of Maine's only fresh strawberry pie competion. Harvey is applauded and cheered. Grinning, she thanks the judges, saying simply that she is motivated by a love of cooking and an adventurous spirit in the kitchen.

Onlookers mosey past the table for one last yearning look at the luscious pies then they drift out the barn into the afternoon sunlight.

Six Blue Ribbon Recipes

This is a first-place winner in the 2010 Union Fair's Two-crust Blueberry Pie Baking Contest

Crust:

2½ cups all-purpose flour
1 teaspoon salt
½ cup shortening
1 stick (½ cup) chilled butter
6-8 tablespoons ice water
1 tablespoon cider vinegar

Filling:

5 cups fresh or frozen wild blueberries
¾ cup sugar
3-4 tablespoons KAF fruit pie thickener
1 tablespoon lemon juice
½ tablespoon cinnamon
pinch of salt

Crust: Place flour and salt in food processor and pulse a few times to combine. Add butter and shortening to the flour mixture and pulse until mixture resembles coarse meal, about 10 seconds. Remove mixture to a bowl. Add the vinegar and water. Mix until the dough holds together. Divide dough in half. Press each half into a flattened circle, wrap in plastic, and refrigerate for at least 30 minutes before rolling.

Filling: Place 1 cup blueberries in heavy saucepan with 1/4 cup water. Mix sugar, salt, and thickener together and mix with the berries in the pot. Cook for a few minutes until sugar is dissolved and mixture is clear. Take off heat. Add lemon juice and cinnamon. Add the other four cups of berries. Stir until well mixed—being gentle so as not to crush the berries. Pour into pastry-lined pie plate. Add top crust, crimping the edges and making steam vents. Bake at 425°F for 15 minutes, then lower heat to 375°F and bake for 30 minutes.

This coffee cake took first place in the 2010 Union Fair's Wild Blueberry Dessert Contest

½ cup of butter
1 cup brown sugar
2 eggs
1 cup sour cream
1 teaspoon vanilla
2 cups flour
1 teaspoon baking powder
½ teaspoon baking soda
¼ teaspoon salt
1 cup Maine wild blueberries

Filling:
⅓ cup chopped nuts
⅓ cup brown sugar
¾ teaspoon cinnamon

Glaze:
¾ cup powdered sugar
1 Tablespoon butter, softened
1 Tablespoon milk
½ teaspoon vanilla

Preheat oven to 350°F. Cream together butter, sugar, eggs, sour cream, vanilla. Then add dry ingredients and fold in blueberries. Put half the batter in a greased bundt pan, then add in the filling. Add in the rest of batter. Bake for 45–50 minutes. Turn out on a plate and let cool. Drizzle with glaze.

These muffins took first place in the 2010 Union Fair's Wild Blueberry Muffin Contest

⅓ cup margarine, softened
⅔ cup sugar
¾ cup milk
2 eggs
2 cups flour
3 teaspoons baking powder
½ teaspoon salt
2 cups fresh blueberries, dusted in flour

Topping:

¼ cup margarine, softened
½ cup brown sugar
1 teaspoon cinnamon
½ cup flour

Mix together margarine, sugar, egg, and milk. In another bowl, combine flour, baking powder, and salt. Add to cream mixture. Fold in floured blueberries. Place in muffin pan. Mix together topping ingredients and sprinkle on top of muffins. Bake at 425°F for 16–20 minutes or until a toothpick comes out clean. Serve warm or cold.

Andy Koch's

Blueberry Spice Whoopie Pies

This is a first place winner in the children's cookie division of the 2010 Machias Wild Blueberry Cooking Contest. Andy also claimed the title of Best Boy Cook

Cake:

2 cups all purpose flour

1 teaspoon ground cinnamon

½ teaspoon ground ginger

¼ teaspoon ground nutmeg

¼ teaspoon salt

¼ teaspoon ground cloves

½ teaspoon baking soda

1½ sticks butter, softened

1 cup packed light brown sugar

1 large egg

2 tablespoons light molasses

½ cup reduced-fat sour cream

Filling:

1 (8-ounce) package reduced fat cream cheese

1 jar marshmallow creme

2 cups blueberries, plus additional for garnish

Andy Koch's

Blueberry Spice Whoopie Pies

To make the cakes: Preheat oven to 350°F. Spray 3 large cookie sheets with nonstick cooking spray. In medium bowl, mix together the flour, cinnamon, ginger, baking soda, nutmeg, salt, and cloves. In large bowl, with mixer on medium-high speed, beat butter until smooth. Add brown sugar and beat 3 to 4 minutes or until creamy, scraping sides of bowl with rubber spatula. Add egg and molasses and beat until well blended. With mixer on low speed, beat in flour mixture, just until blended. Spoon batter by heaping tablespoons onto prepared cookie sheets, 2½ inches apart. Bake cakes one sheet at a time, 11 to 13 minutes or until cake springs back when pressed lightly. Cool cakes in pan on wire rack 1 minute, then transfer with spatula to wire rack to cool completely. Cakes can be made ahead up to one day; wrap tightly in plastic wrap or foil and store at room temperature.

To make the filling: In large bowl, with mixer on medium-high speed, beat cream cheese until smooth. Reduce speed to low; add marshmallow crème and beat just until blended, scraping beaters if necessary. Fold in blueberries. Spread ¼ cup filling on flat side of half the cakes. Top each with plain cake, flat side down, pressing lightly. If you like, garnish with additional blueberries pressed into the sides.

Not only is Ellen Farnsworth the chairwoman of the Machias Wild Blueberry Festival, she makes a mean salsa. This is her prize-winning recipe

2 medium UglyRipe tomatoes, peeled, seeded, and diced
1 cup wild blueberries
¾ cup finely chopped onion
2 cloves garlic, minced
2 tablespoons olive oil
2 tablespoons rice vinegar
1 jalapeño pepper, finely chopped
¼ finely chopped Hungarian wax pepper
2 tablespoons chopped fresh cilantro
2 tablespoons chopped fresh flat parsley
salt and pepper to taste

Mix well and serve.

This is a prize-winning salad from the Machias Wild Blueberry Festival.

6 medium-sized golden beets
1 small fennel bulb, cleaned and thinly sliced,frond reserved
2 cups wild blueberries
1 large lemon, juiced
2 tablespoons olive oil, divided
salt and pepper to taste

Dressing:
1 cup balsamic vinegar
½ cup wild blueberries
2 tablespoons finely diced shallot

Preheat oven to 400°F. Peel beets and dice into 1/2-inch cubes. In a mixing bowl, coat the diced beets with 1 tablespoon olive oil and season with salt and pepper. Spread on baking sheet and roast 15 to 20 minutes or until tender. Combine the fennel and beets in mixing bowl and dress with lemon juice. Carefully stir 2 cups of blueberries into beets and fennel. Add salt and pepper to taste. Store in refrigerator at least 2 hours and up to one day.

To make the dressing: Over medium heat, add 1 tablespoon olive oil to a medium saucepan. Add the shallots. Cook until softened. Add 1 cup of balsamic vinegar to saucepan. Simmer until reduced by one-quarter of its original volume. Remove vinegar and shallot mixture from heat and stir in 1/2 cup of the blueberries. Dress the chilled salad with the warm blueberry vinaigrette. Serve with torn fennel frond as garnish.

After the Harvest

Inside the Packing Plant

WINNOW

Within 24 hours of harvest, the blueberries are delivered to the packer. These facilities vary in size, from small barn operations like Del Emerson's in Addison to large factories like G.M. Allen & Son, Inc. in Orland and Jasper Wyman & Sons in Cherryfield and Deblois. The berries are poured into a winnowing machine, which blows out the leaves and twigs.

REMOVE ROCKS

The berries emerge from a winnower onto a rock eliminator, a large vibrating sifter that allows pebbles and debris to drop away.

A worker operates the rock eliminator at Jasper Wyman & Son in Deblois.

FRESH PACK

FRESH PACK OR PROCESS?

FRESH PACK

Berries destined for "fresh pack" treatment are harvested with a gentler hand because the fruit is going directly to stores where it may sit for four or five days. The berries cannot be wet when raked, and the raker must take care to avoid tearing the skin, which attracts fruit flies and hastens rotting. The harvesting crates are generally filled only halfway so the berries aren't crushed.

SORT

The rock eliminator moves the blueberries to the tilt belt. Its angled bed allows firm blueberries to roll down to the next part of the packing line. Soft berries, sticks, and leaves don't roll as easily, so they stay on the tilt belt and are conveyed off the end.

HAND PICK

Packing line workers stand on either side of a conveyor belt, picking out any unripe and damaged berries and other debris that the machines failed to eliminate.

PACK

A worker catches the sorted blueberries in boxes as they fall off the end of the belt. Because they have not been rinsed, fresh-pack blueberries retain the distinctive dusty white bloom that is their natural protection against the sun.

EAT SOON OR LATER?

EAT SOON

DELIVER

Boxes of fresh berries are trucked to stores. They represent less than 1 percent of Maine's blueberry harvest.

LATER

FREEZE

The boxes are stored in a freezer at –15°F.

PROCESS

(Next Page)

PROCESS

Because of their fragile nature, wild blueberries are processed for long-distance travel. Ninety-nine percent of the berries are preserved by freezing, though some of these will be canned later. Seven companies operate blueberry processing plants in Maine.

WASH

The berries drop from the rock eliminator into a shallow bath. Green and red berries float to the top, while the blue fruit sinks to the next packing line section. Imperfect berries will be used for juice. At Jasper Wyman & Son, this step is repeated.

COOL and DRY

The berries move through a sanitizing spray of cold water that brings down their temperature in preparation for freezing. A vacuum drier pulls out the excess water.

FLASH FREEZE

The blueberries are conveyed into the freezer tunnel, which is roughly the size and shape of a tank truck. The temperature inside: -20 to –40°F. At the end of the tunnel, a spinning cylindrical mesh cage eliminates stems and imperfect berries to be used for juice stock. Each berry's journey through the tunnel lasts about thirteen minutes.

COLOR SORT

An Elbiscan laser color sorter scans frozen berries, removing anything that isn't bluish purple and shuttling it into a bin for juice stock.

HAND PICK

The frozen berries get one last inspection, this time by workers who pick out any that are imperfect.

PACK and STORE

The berries are collected in bins, to be weighed and boxed. They are stored in refrigerated warehouses until they are trucked to market.

Boxes and boxes of blueberries in refrigerated storage at GM Allen & Sons, Inc.

Opposite page: Berries being washed and cooled down before entering the freezer tunnel at Jasper Wyman & Son

Down East Meets South Asia

Used to be, blueberry products were pretty much limited to jam and syrup, but the quadrupling of the blueberry harvest over the past two decades, combined with the seemingly nonstop good news about the health benefits of blueberries, has spawned a creative explosion of blueberry goods. Indeed, a few small wineries, like Savage Oakes Vineyard & Winery in Union and Bartlett Maine Estate Winery in Gouldsboro, were born out of lowbush blueberry fields, and they in turn have inspired the addition of wild blueberries to soda, beer, even coffee. There's chocolate-covered blueberries (and whoever invented them deserves a prize), dried blueberries (sort of like raisins), blueberry ice cream, blueberry dog biscuits, and a slew of personal hygiene products like blueberry soap, lip balm, and shampoo.

Many in the new generation of blueberry entrepreneurs have invented their products out of twin passions. Such is the case for Molly Sholes, who as a young bride in the mid-1950s settled with her husband, an employee of the United States Information Agency, in Lahore, Pakistan. Assignments over the next twenty-seven years would take them to Bombay, New Delhi, and Madras. Molly came to love the cuisine, which is distinguished by its aromatic spices. Friends and professional chefs taught her to cook many flavorful regional dishes, and she even learned to roast and grind her own garam masala, a commonly used pungent blend of spices.

The Sholeses spent their holidays in Maine, where they had a house sitting among thirty-eight acres of blueberry fields on Spruce Mountain in West Rockport. "I began to learn the techniques of growing and harvesting low bush blueberries," Sholes writes in her self-published cookbook, *East Meets West*. "The farm had no electricity, so we brought rice-winnowing baskets from India to cull good berries from bad, and then sold fresh berries locally."

When the couple retired to Spruce Mountain, Molly combined her love for Indian cuisine with that of Maine blueberries, beginning with a spicy blueberry condiment that she adapted from a plum chutney. Today wild blueberry chutney and wild blueberry chutney with almonds and raisins are among several products that Molly prepares in her own kitchen and sells under the Spruce Mountain Blueberries label.

Molly Sholes' Blueberry Chutney Chicken

Serve with green peas over basmati rice or fettucine

2 cloves garlic, chopped
1-inch-square piece of fresh ginger root, peeled and diced
1 bunch scallions, coarsely chopped (including green tops)
4 tablespoons unsalted butter or vegetable oil
1½ pounds boneless, skinless chicken breasts
⅓ cup white vermouth
½ cup chicken stock
⅓ cup sour cream
2 fresh green chilies, chopped (or about 2 tablespoons canned green chilies)
½ cup wild blueberry chutney
salt and pepper to taste

In a heavy skillet, saute the garlic, ginger, and scallions in butter or oil. Salt and pepper the chicken breasts and brown in the same skillet (this takes only a few minutes). Remove the chicken breasts to a baking dish and swirl the vermouth and stock into the skillet, blending them with the juices and bits of garlic, scallions, and ginger. Remove the skillet from the heat and add the sour cream, chopped green chilies, and chutney to the pan juices. Pour the sauce over the chicken breasts in the baking dish. Bake, covered, at 350°F for 20 to 25 minutes. Serves 4–6.

From the grower scorching shrubs to the worker swinging his rake in a sun-baked field to the line picker painstakingly sorting good berries from bad, blueberries are a way of life in Maine. So next time you pick them yourself, seek them out at a farm stand, or buy them frozen from the supermarket, consider all the tradition packed into those beautiful little blue orbs. Then, for goodness sake, pop them in your mouth and enjoy.